YOUR KNOWLEDGE HAS VALUE

- We will publish your bachelor's and master's thesis, essays and papers

- Your own eBook and book - sold worldwide in all relevant shops

- Earn money with each sale

Upload your text at www.GRIN.com and publish for free

Bibliographic information published by the German National Library:

The German National Library lists this publication in the National Bibliography; detailed bibliographic data are available on the Internet at http://dnb.dnb.de .

Imprint:

Copyright © 2016 GRIN Verlag, Open Publishing GmbH
Print and binding: Books on Demand GmbH, Norderstedt Germany
ISBN: 978-3-668-19730-5

This book at GRIN:

http://www.grin.com/en/e-book/319352/agricultural-production-in-nigeria-a-review-of-recent-developments-in

Seun Kolade

Agricultural Production In Nigeria. A Review of Recent Developments in Policy and Practice

GRIN Publishing

GRIN - Your knowledge has value

Since its foundation in 1998, GRIN has specialized in publishing academic texts by students, college teachers and other academics as e-book and printed book. The website www.grin.com is an ideal platform for presenting term papers, final papers, scientific essays, dissertations and specialist books.

Visit us on the internet:

http://www.grin.com/

http://www.facebook.com/grincom

http://www.twitter.com/grin_com

Table of Contents

Agricultural Production In Nigeria: A Review of Recent Developments in Policy and Practice

By Seun Kolade, PhD

International Development, Emergencies and Refugees Studies Research Group,

Department of Social Sciences, London South Bank University.

Abstract

This article draws on critical review of research articles, official statistics and policy documents on agricultural production in Nigeria, especially since the beginning of the current democratic dispensation in 1999. This paper highlights the findings of researchers and policy makers on the significant gains made in the agricultural sector, especially in terms of increased productivity of staple crops like cassava and rice. However, this review also identifies significant gaps in knowledge and deficiencies in practice, in the areas of innovation diffusion among rural farmers, market reforms, engagement in high value chains, and the politics of policy implementation and evaluation.

Keywords: Agricultural production; rural development; innovations; food security, Nigeria

1. Introduction

Following the end of a 16-year military rule and the commencement of the fourth republic in 1999, there has been renewed interest and progress in the agricultural sector in Nigeria. The country has now experienced, for the first time, 15 years of uninterrupted civilian rule. This paper brings together a wide range of literature on agricultural production in Nigeria, with focus on academic research, official statistics and policy documents. The aim of this is to highlight key findings in research and gaps in knowledge, and provide critical reflection on the main issues and challenges associated with the design and implementation of official intervention efforts on food security and rural development. Thus, there is a discussion of farming practices in Nigeria, with particular emphasis on the impacts of seasonal farming, environmental degradation and continued reliance on traditional implements and methods. This is followed by a critique of the social organisation of agricultural production, with focus on issues relating to mobilisation of labour, gender, and land tenure. Finally, there is a discussion of challenges associated with agricultural credit and appropriate technology, and how policy interventions of successive governments have sought, or failed, to address those key challenges.

The body of secondary data referenced in this paper has been carefully selected from established sources, including the Nigerian National Bureau of Statistics (NBS), Food and Agriculture Organisation of the United Nations (FAO) and the World Bank, among others. Where possible, the secondary data from these organizational sources have been triangulated with empirical data from scholars whose works have been cited in this review, to enhance reliability and validity.

1.1 Land use distribution

Nigeria, with an estimated population of 170 million people, is the most populous country in Africa, and about 50% of the population live in the rural areas. The total land area is 911,000sq.km, and about 80% of this is available for various agricultural purposes, including arable land, permanent crops, pastures, and irrigated land, as shown in fig.3.1 (FAO, 2013).

Although Nigeria is the world's largest producer of cassava, yam and cowpea, it is still a food deficit nation and a net importer, and more than 80% of rural dwellers in particular live below the poverty line. Agricultural land is severely underutilized, with less than 50% of land cultivated as of 2009, and a very small fraction of irrigable land has been irrigated (IFAD, 2009).

As table 1a shows, production of staple food crops like yam and cassava has generally increased in the past ten years, while production of cash crops like oil palm and groundnut has decreased or stagnated (FAO, 2013). Cocoa production, in particular, reduced by 25% from a decade peak of 485,000 tonnes in 2006 to 363,510 tonnes in 2009. Based on constant 2004-2006 1000 1$ (1000 Int. $), the total value of agricultural production between 2003 and 2011 has moved between Int.$31 billion to Int.$36 billion, the peak coming in 2008 (table 1b). In other words, there has not been significant or steady upward trend in the value of agricultural produce in the past decade.

The contribution of agriculture to GDP has been put at 23%, and about 90% of food and agricultural output is from small-scale farmers who cultivate between 0.5Ha in the densely populated region to about 4 Ha in sparsely populated areas. Among other things, neglect of rural infrastructure, land degradation and drought, and non-availability and /or non-access to equipment and innovations have resulted in low yield and non-profitability of agricultural production (FAO, 2013; IFAD, 2009).

From the foregoing, the role of small-holder farmers in agricultural production is a critical focus of inquiry, and, with respect to this, the impact of technological innovations in mitigating vulnerabilities and enhancing productivity and profitability.

2. Rural poverty, food poverty and farming practices

Statistics on urban/rural poverty in Nigeria indicate that, especially from 1980 onwards, more and more people are now living below the poverty line, and rural dwellers are significantly affected much more than urban dwellers (Omonona, 2010; World Bank, 2013). As shown in fig. 2, in 1980, 28% of rural dwellers, compared with 17% of urban dwellers, lived below the poverty line. By 1985, the poverty figures for urban dwellers had risen sharply to 38%, while rural poverty significantly increased from 28% to 51% within the same period. By 1996, urban poverty had risen to 59%, while rural poverty skyrocketed to 72%.

2.1 Causes of rural poverty

In addition to the challenge of population explosion, a major reason for increasing levels of poverty in Nigeria is the disproportionate allocation of resources by successive governments and consequent wide gap and disparity in pace of development in the rural areas compared with urban centres (Ogunlela & Ogunbile, 2006). This trend has, on another level, played itself out with regard to an increase in urban poverty too, as more and more resources and economic were concentrated and appropriated in the hand of a minority of political elite. For example, following independence in 1960, many urban centres began to enjoy infrastructures like roads, water supply and educational and social amenities, but the vast majority of rural communities, where more than 60% of Nigerians lived, were almost entirely neglected. Rural farmers struggled greatly to transport their produce to large markets in the urban areas because of poor or non-existing roads. In addition, farmers in rural areas lack access or have restricted access to land and credit, and continued to rely on primitive farming implements in the face of more challenging environmental conditions and depletion of soil resources. As a result, agricultural production suffered greatly, along with household incomes (Apata et al, 2010; Omonona, 2010; Ogunlela & Ogunbile, 2006; Kolade, Harpham, & Kibreab, 2014).

Farming practices in Nigeria

One of the factors associated with the prevalence and continued increase in rural poverty is the state and quality of farming practices in Nigeria. Some surveys indicate that about 44% of male farmers and 72% of female farmers cultivate less than 1 hectare per household, despite the fact that about 34million ha out of 83million ha – 40.96%- of agricultural land is currently being cultivated. Reports also indicate that about 90% of Nigeria's food is produced by small-scale farmers cultivating small pieces of land (IFAD, 2009a; IFAD, 2009b). Moreover, for the less than 50% of agricultural land currently being cultivated, yield per hectare is low compared to other developing countries like Brazil and Thailand, owing to the following main factors:

i)   Prevalence of rain-fed agriculture: the vast majority of farmers in Nigeria depend mainly, or only, on seasonal rainfall for their cultivation, such that agricultural production is suspended for about half of the year when it does not rain (Nnamchi & Ozor, 2009; Nzeadibe et al, 2011). A very small fraction of irrigable land, less than 1% according to one estimate, has been irrigated (FAO, 2013), at considerably higher cost compared with other African countries.

ii)   Impact of environmental degradation: climate change and land degradation are major contributors to low yield and crop failures (Nzeadibe et al, 2014), and the impact is especially high in Africa, where it is projected that crop yield may fall by between 10 to 20%, or even up to 50% in some cases, by the year 2050 (Anete & Amusa, 2010). Among others, climate change Nigeria has brought about significant seasonal changes in rainfall and temperature, drought, alteration in evapo-transpiration, photosynthesis, and biomass production, and negative impact on land suitability for agricultural cultivation (Medugu, 2009; Obiora, 2014). In the Niger Delta, in particular, the challenge of climate change is aggravated by very high rate of gas flaring (Akinro et al, 2008). Climate change is considered probably the most serious environmental threat to the fight against hunger, malnutrition and disease in Africa (Anete & Amusa, 2010).

iii)   Continued use of primitive implements and methods: Nigerian traditional agriculture is dominated by the use of cutlass and hoe, implements which are used by the vast majority of rural farmers (Nkanini et al, 2006; Oluyole, et al, 2013). In particular, the traditional farming hoe is a major tool of Nigerian agriculture. It is usually made of wooden handle attached with metal blade forged by local blacksmiths, although basic variations in types and sizes exist across the geographical and ethnic zones. The hand-hoe, like the cutlass and machete, is high in energy demand and low in productivity. The main alternative to this is the 'animal draught power farming', which entails the use of work bulls capable of supplying about one-tenth of their body weights working continuously, compared with 0.07KW for direct human labour (Sanni, 2008; Oni, 2011). However, animals can work only for limited duration per day, and the feed requirements for the working animals also make it sometimes too expensive and difficult to be employed by poorer farmers. A recent study indicate that only 1% of farm power is supplied by mechanical power in Nigeria, followed by 10% for animal drawn systems, and 89% for direct human labour (Oni, 2011; FAO, 2005). In addition, practices like bush burning is used as the main method of clearing farm land, thereby increasing the amount of greenhouse gases in the atmosphere, and effectively aggravating the problem of global warming and climate change (Anete & Amusa, 2010).

The examples of cassava and rice

The impacts of factors listed in the foregoing can be examined with specific examples of cassava and rice, two main stable food crops and potential sources of foreign exchange earnings for Nigeria. Cassava production in Nigeria is the largest in the world, with a 2002 FAO report indicating production was approximately 34million tonnes, a third more than Brazil, the 2nd largest producer of Cassava in the world (FAO, 2005). Cassava production received a significant boost between 1988 and 1992, following the release of improved varieties by International Institute of Tropical Agriculture (IITA). In 2002, a presidential target was set to achieve $5 billion in value-added foreign exchange earnings from cassava production, requiring an expansion of cultivated area from 3mil to 5mil ha, and an increase in yield per ha from about 10 to average of 30 tonnes per ha, in order to achieve a total production target of 150mil tonnes by the end of 2006, up from 34mil tonnes in 2002. More conservative estimates for a production target of 60mil tonnes, suggested by some research institutes, require an expansion of cultivated area from 3mil to 4mil ha, and a yield increase from 10 to 15tonnes per ha. However, majority of the farmers were not incorporated into the presidential initiative (Awoyinka, 2009), and recent data show that cassava production was at

44.5mil tonnes in 2008 and had only increased by 3 mil tonnes between 2005 and 2008. Yield per ha remained stagnant at about 10 tonnes per ha, and there was little expansion in cultivated area (IFAD 2009; FAO 2013; Philips et al, 2004). In effect, the potentials, even with conservative estimates, have not been achieved, mainly because the use of mechanised farming and improved varieties, among other better farming practices, have not increased proportionately to meet yield requirements. Moreover, value-added activities, like processing and storage, are severely under-developed for cassava and other crops in Nigeria (Philips et al, 2004; Sanni et al, 2005; Adebayo, 2009).

Similarly, with respect to rice production, Nigeria is the largest producer in West Africa, with production estimated at 3.28mil tonnes in 1999, and the 2007 and 2008 showing fluctuating figures of 2.99mil tonnes and 4.17mil respectively. Nevertheless, using the 1999 figures, Nigeria was also the largest importer of rice in West Africa, with importation estimated at 1mil tonnes in 1998, at the cost of about $300mil (FAO, 2013). It is also estimated that about 2.2mil ha of land was cultivated for rice production in 1999, compared with 5mil ha available, and average yield was, and remains, 1.5tonnes per ha. In comparison, potential yield for rice is about 6 tonnes per ha, especially in the lowland which constitute about 65% of land cultivated for rice. This potential is not met mainly on account of poor water control system, bad soil management, shortage of input, and annual flooding, in addition to insignificant increase in the area of land cultivated for rice (EDO, 2003; Fashola, et al, 2005).

This paper now examine the systems and processes by which labour is deployed and land other resources utilized for farming activities in Nigeria. These include the role and impact of government policies, at national and local levels, on management and allocation of resources.

3. Social organisation of agricultural production

Before the advent of colonial administration, traditional societies and ethnic groups in different parts of Nigeria had an elaborate and fairly organized process by which they harnessed different factors of production, especially land and labour, for agricultural production. The underlying principles including communal land ownership and labour sharing, applied in various degrees with slight modifications, are essentially similar across the whole spectrum of ethnic groupings. One of these ethnic groups, the Kofyar people of central Nigeria, was the subject of a classic work by Robert Netting in the 1960s. The general thrust of the findings is representative of what obtained in many rural parts of the country, and some of the practices are still observed today (Netting, 1968; Stone, 1998).

3.1 Mobilisation and organisation of labour

As in many African societies, the household is the primary unit of labour mobilization for agricultural production. Typically, the head of the household is the man, although in practice, and as will be explored in more detail later, women exercise considerable independence and control. Children are also seen as important sources of labour at the household unit, and this notion is often associated with the practice of polygamy, especially in the rural areas (Oluyole et al, 2013). However, the value and productivity of household labour resources depends significantly on the age and health of the farmer, and whether or not the children are young and dependent or older enough to fully participate in farm work (Olaniyi et al, 2007).

Requirements for labour often vary with respect to seasons, type of crops and type of farm lands. It is suggested, for example, that cultivation in the savannah is more demanding compared with forest lands, and post-harvest labour requirements are more intensive for some crop, compared with others (Oluyole et al, 2013). In addition, there is usually shortage of labour at seasonal peaks in the rainy season when much of the cultivation is done. The dry season also has its own peculiar problems- for crops planted at this period- due to the challenge of labour migration to non-agricultural engagements (Stone, 1998; Abila, 2012). However, the impact of rural-urban migration on farm labour availability has not been given adequate attention in the literature, considering the long history of neglect of rural infrastructure, and disproportionate attention of governments on urban areas.

These challenges with labour shortages are mitigated by a variety of labour sharing strategies and methods, as exemplified by, but not exclusive to, the Kofyar people. In addition to mobilization of household labour, which is flexible and readily available, the Kofyar employed two other strategies of labour mobilization: the *Mar-mous* labour parties, and the *Wuk* exchange-labour groups. The *Mar-mous,* which has to be arranged weeks in advance, typically involve large neighbourhood work groups, usually comprising of between 30 to 60 workers but can sometimes exceed 100. Millet beer is served after work, which normally lasts from morning to late afternoon. The *Wuk* groups are usually smaller in size- between 5 and 20 workers- and are essentially voluntary associations whose members work in turn on each others' farms. This reciprocal labour is effectively used as payment for the workers, and carefully documented (Stone, 1998).

The benefits and advantages of these labour sharing strategies include avoidance of high cost of hired labour, better quality of labour where particular skills are required, enhanced output and productivity in group labour especially in a festive environment like *Mar-mous*, and comparative effectiveness and efficiency of group labour where simultaneous labour is required, as is the case for harvesting and storage of millet (Netting, 1968; Stone, 1998).

The fundamental weakness of these traditional methods of labour mobilisation is that the output and efficiency is very low. As stated earlier, there is the need for wider investment in, and diffusion of, labour saving technological innovations (Abila, 2012; Oluyole et al, 2013).

3.2 Ownership and utilization of land

The land tenure system in Nigeria, by which individuals gain access to land for agricultural, residential and commercial purposes, varies in different part of the country, but the communal system of land ownership, by which control and allocation is vested on groups and kinship, is the most prevalent among most ethnic groups in Nigeria. Under this system, large parcels of land are owned by clans and extended families, and shared among members of the group. Chief and village heads, with proportionately bigger holdings on account of their status, also play important roles in the control of access to lands (Philip, et al 2009; Onyebinama, 2005). Under this arrangement, much of the land so acquired is transferred across generations by means of inheritance.

The communal system of land ownership has its own merits and potential advantages, especially with regard to the fundamental fact that land is shared among all recognised members of the community, who belong to one or the other family. However, the system also presents significant difficulties and bottlenecks that hinder the allocation and distribution of land for agricultural purposes. First, the sharing of lands among, as well as within, families, is not necessarily equitable, such that some individuals tend to own significantly more lands than others, on account of their status. Also, the fragmentation of lands, beyond the basic requirement for shelter, means that large portions of agricultural lands are left unused because owners are not interested in farming, and tenure systems and conditions for agricultural users are not secure enough for continuous, long-term cultivation. Finally, the access rights of individuals outside the owning group or family is severely restricted, such that those individuals are unable to fully use the lands as collateral for agricultural credits (Economic Commission for Africa (ECA), 2004; Onyebinama, 2004; Philip et al, 2009).

In 1978, in a bid to correct some of the weaknesses and difficulties inherent in the communal system of land ownership, the military government promulgated the Land Use Act (LUA), later updated in the 1990 version. This law transferred the ownership and control of lands to the states and local governments, under the leadership of the governors and chairmen respectively. The state government was vested with direct control of lands in urban areas, while local governments exercise control over allocation of lands in the rural areas, with the exception of lands in excess of 500 ha allocated for agricultural purposes, and those in excess of 5,000 allocated for grazing purposes, both of which require the state governor's consent (Federal Republic of Nigeria, 1978; Olayiwola & Adeleye, 2006; Fajemirokun, 2002)

The LUA achieved some of its main objectives, but in other respects, it simply replaced or aggravated the difficulties and bottlenecks of the traditional system with similar ones in which government authorities often abused the allocation of lands placed in their trust. Moreover, a good percentage of the lands, especially in the rural areas, are still owned by communities under customary laws, and controlled by members to whom they are allocated, with all attendant benefits and problems (Philip et al, 2009). There is need for further research on why past and current interventions have failed, and what can be done in the future, to improve access of small—scale farmers to agricultural lands.

3.3 Gender issues in agricultural production

Whereas men control most of the resources for agriculture, and take predominant role in decision making processes, women constitute a very strong percentage of the work force for agricultural production. A recent study showed that women constitute between 60 and 80% of agricultural labour force in Nigeria, producing about two-thirds of staple foods (Ogunlela & Mukthar, 2009), although other studies have put this figure at less than 50% (Audu, 2009; Mohammed & Abdulquadri, 2012). In some states, it is reported that women are almost entirely responsible for the production and processing of arable crops (Ukeje, 2004; Idowu & Ajani, 2008).

It is noted, however, that there are certain variations in different parts of the country regarding the nature and extent of women's participation in agriculture, based on religious and cultural assumptions and attitudes. In some communities in the core north, like Zaria, it is reported that Muslim women were mostly in domestic seclusion and did little farm work outside, but helped with some storage and processing activities-like groundnut shelling- in the home compound. In other parts, like South East Nigeria, farm and crop specializations were determined according to gender (Audu, 2009). Nevertheless, in allocation of rewards and wages for labour women get less than men, obtaining an average of 75% of men's wage for one day of agricultural labour (Idowu & Ajani, 2008).

With respect to participation in decision-making, a recent study conducted in the Northern Zaria state, reinforcing the general trend in the country, showed that, for most parts of decisions regarding land purchase, land preparation, time for sowing, weeding, harvesting and other aspects of farming activities, consultation with women farmers is less than 20%, and they are almost entirely ignored in making final decisions (Ogunlela & Mukthar, 2009). In particular, poorer and less educated women are most neglected in decision making (Damisa & Yohanna, 2007). The situation has improved slightly in recent years, with the advent of intervention programmes such as World Bank -supported Women in Agriculture (WIA), designed to bridge the gap in extension services to women, and the efforts of such groups like Women Farmers Advancement Network, an umbrella organisation working with more than 250 women groups in several states of Northern Nigeria (Sabo, 2006; Ogunlela &

Mukthar, 2009; Idowu & Ajani, 2008). These intervention programmes have been generally limited by inadequate attention to the relevant needs of the women, and lack of access to land, among others (Odurukwe et al, 2006).

In general, agricultural production in Nigeria faces two major challenges, as observed in the foregoing. One is the problem associated with non-availability, or severe restrictions, of agricultural credit which enables farmers to access and apply various inputs for improved production and profit. The other concerns the various difficulties encountered with availability, cost and relevance of appropriate technology necessary for improved yields. These two challenges, which are intertwined, will now be considered in some detail in the following section.

4. The challenge of agricultural credit and appropriate technology

The history of government interventions in agricultural production in Nigeria has rightly been characterised by focus on two important elements: provision of credit and promotion of technological innovations. Funds are required for various investments necessary for production, processing and marketing of agricultural products, and technology is essential for improved productivity and profit. A summary of government intervention programmes and schemes is given in the next section, but the current section examines the challenges and opportunities for agricultural financing and technology application.

4.1 Agricultural financing

Traditionally, small-scale farmers, especially in the rural areas have mostly relied on informal means of financing, including loans from family members and friends who lend to farmers more as a social obligation. In addition, produce buyers lend money to farmers in anticipation of harvest, as is sometimes the case with cocoa farmers. Finally, there is the *esusu* credit arrangement, by which members contribute certain sums of money and hand the total amount to each member in turn until all members have had their share, making large sums available to farmers when they need them (Oloyede, 2008; Iganiga & Asemota, 2008). However, the informal means of credit acquisition is fraught with various difficulties and uncertainties. Funds from friends and family members are essentially discretionary, and are unlikely to be sufficient or regular, partly because there is no commercial incentive and expectation of profits, and family members are often of similar economic status to borrowers, with limited funds available. There are also challenges associated with inadequate book-keeping system, and absence of standardised laws and framework (Oloyede, 2008). Similarly, funds from produce buyers require, especially with respect to the size of funds needed, considerable guarantee regarding the ability of the farmer to pay at harvest. The *esusu* arrangement appears to be the best of all, but also require lots of trust and effective organization, as well capacity to deal with failures of members to pay.

In the light of the foregoing, and in recognition of the critical importance of credit availability for agricultural production, the federal government of Nigeria has undertaken several intervention initiatives over the years since independence. Among the reasons given by the government of Nigeria for various interventions in agricultural financing are: i) to smoothen out the imperfections in the agricultural financial market; ii) ensure food security by credit provision to farmers and direct production; iii) achieve favourable balance of payments by facilitating less-reliance on food imports; iv) promote foreign exchange earnings; v) enhance other social-economic factors like poverty reduction, employment generation, food price stability and reduction in rural-urban migration; and vi) use finance as engine of growth and development and encouraging farmers by providing concessionary loan terms (Eze et al, 2010). The aims of the intervention programmes do not appear to give necessary attention to strengthening and integrating the informal, traditional systems of financing with formal systems.

The credit intervention schemes launched by the government over the years include the followings: i) The Agricultural Credit Guarantee Scheme Fund (ACGSF), established in 1978, which provides 75% guarantee backed by the central bank on agricultural credit in default, and is targeted mainly at small-scale farmers; ii) Small and Medium Enterprises Equity Investment Scheme (SMEEIS), founded in 2001 to provide financing in form of equity or debt for small scale and medium enterprises, including agro and agro-allied businesses; iii) Refinancing and Rediscounting Facility (RRF) established in 2002 to provide liquidity for banks that lend long-term to agriculture; iv)Agricultural Credit Support Scheme (ACSS), set up in 2006 with contributions mainly from CBN to finance large agricultural projects; v) Large Scale Agricultural Credit Scheme (LASACS), established in 2009 with a fund of N200 billion to finance large integrated farm projects in the wake of the global economic crisis; vi) Supervised Agricultural Loan Board, set up by state governments to dispense loans in form of credit to farmers (Eze et al, 2010; Adetiloye, 2012).

In spite of the grand and elegant ambitions on paper, actual commitment of funds by government has not met up the projections and aspirations, in terms of budgetary allocations and administration of funds. A study on budgetary allocations between 1990 and 2002, indicate that the percentage of total national budget allocated to agriculture 1.28 % in 1999, and typically fluctuate between 1.7 and 4. 9%. The only exception is 2001, when the budget allocation reached a peak of 6.8% (Eze et al, 2010). In addition, farmers encounter all sorts of administrative and bureaucratic bottlenecks, exacerbated by the difficulty inherent in administering loans to rural farmers scattered over vast areas covered by long distance (Iganiga & Asemota, 2012; Eze et al, 2010).

4.2 Technology for agricultural production

The impact of technology in agricultural production is an important consideration in a discussion about building capacity, improving yield, and expanding areas used for farming activities. Some emphasis is here required on the character and function of low-input technologies, employed by the vast majority of small-scale farmers in Nigeria. Also, it is useful to compare and evaluate potentials and opportunities for increased efficiency of mechanised and non-mechanised farming, as well as explore how technology can help in the challenge of coping with climate change, and how extension service and effective input marketing and distribution, help in disseminating information and promoting the application of technological innovations.

i) Low-input technologies: these are based mainly on refinement of indigenous knowledge and practices. In Nigeria, such technologies include, among others, use of ruminant farm animals for clearing of bushes and residues of harvested stalk, and use of cattle manure, green manure, mulching, urine, urine-marine slurry and manure tea technology for maintaining or enhancing soil fertility. In addition, there are also cereals and pulses low input pest control and storage devices, use of low input bio-pesticides, stock damping technology and yam-mini set technology for seed multiplication, use of non-electricity-based waterbed incubator for egg- hatching, basin irrigation, border irrigation and furrow irrigation methods, hay- making and silage making for animal feed preparation, and low-input power machines like solar power, wind-mills, bicycles, wheelbarrows, draught animals and hand-operated shelling machines (Mpkado & Onuoha, 2008; Sanni 2008)

ii) Comparative efficiency of low-input and high-input technologies: High external input technologies (HEIT), like use of inorganic fertilizers, high-yielding varieties, modern irrigation methods and use of highly mechanised farming have been known to contribute significantly to increase in farm productivity, but they typically require high capital investment which are not easily accessible to small- scale farmers, and potential profit is limited by economies of scale. Conversely, low input technologies, more readily accessible, are usually labour intensive but are considered more sustainable as they emphasize integration of natural processes such as nutrient cycling, nitrogen fixation, and soil regeneration, among others (Anyanwu & Adesope 2010; Tripp 2006). An investigation of non-mechanised maize farms in southern Nigeria indicate that there is more potential- of 38%- for increasing efficiency of farm outputs, compared with 28% for mechanized farming, and another study in northern Nigeria revealed that full use of animal traction technology can bring about up to 78% increase in gross margins (Ajao et al, 2005; Olaoye & Rotimi 2010; Sanni 2008). Also, particularly for high-input mechanised farming, other investigations revealed that efficiency and productivity of technology are hindered by poor operational and maintenance practices by operators, lack of favourable conditions for full integration of agricultural mechanization and lack of essential infrastructure (Davies et al, 2008). More research and funding commitments are thus needed for improvement of low-input technologies, and dissemination of technical and operational information by extension workers and other relevant agents.

iii) Extension service: Access to information and technical knowledge is an important factor in effective adoption and efficient application of technological innovations for agricultural production, and studies have shown that there is positive correlation between frequency of contact with extension workers and farmers increased and sustained use of technological innovations (Omobolanle, 2008). However, the influence and reach of extension workers has been generally modest, and it is suggested that this is partly due to the fact that agricultural extension service in Nigeria is essentially oriented in a top-bottom approach with a focus on training and visit to individual farmers to facilitate technology transfer. An alternative model has thus been recommended, with more emphasis on group approach to enhance better use of limited resources, as well as more participatory approach which draws on local indigenous knowledge and encourages collective ownership of technology. In addition, there is need for more funding commitment on the part of government, decentralisation of extension service to local levels of government, and increased staffing and continuous training of staff (Koyenikan,2008; Akinnagbe & Ajayi, 2010).

iv) Input marketing and distribution: The role of input supply is allied to that of extension workers. In general, extension service provides farmers with useful information about new techniques and methods, including inputs, but it is the input supplier that gets the inputs in the hands of farmers. Nigerian farmers, especially in the rural areas, encounter significant difficulties is purchasing inputs at reasonable, affordable costs, even after they have obtained information about inputs, and have chosen to adopt them (EDO, 2004). This difficulty often arises as a result of geographical distance of farmers from suppliers. In this regard, groups, including formal cooperatives and households with more agricultural links in the social network, fare much better than individuals, as they are able to mitigate costs of travel by buying in bulk. Conversely, suppliers also suffer from the underdeveloped input market. A 2005 pilot project by the International Fertilizer Development Centre (IFDC) showed that 500 producers and dealers, linked up with 750,000 farmers, were able to achieve 38% growth in their businesses, and import and distribution of fertilizers in the private sector increased by 70% for three years running (IFDC, 2005).

5. Agricultural development interventions

Several intervention policies have been spearheaded by successive governments of Nigeria from independence in 1960 to foster and accelerate development in the agricultural sector. Some agencies were also established by the governments, and by non-governmental bodies like the Food and Agricultural Organisation (FAO) of the United Nations to work directly with farmers for enhancement of agricultural production. A summary of these past intervention programmes is set out in the table 2.

The current Nigerian government has embarked on Agricultural Transformation Agenda, as part of the Economic Transformation Agenda, originally designed to run between 2011 and 2015 (National Planning Commission, Federal Government of Nigeria, 2011). According to officials, the current intervention programme was, among other things, designed to place emphasis on agriculture as a business, profit oriented enterprise run by small-scale farmers, rather than a development programme maintained by the federal government (Federal Ministry of Agriculture, 2012; Adesina, 2014). Thus, there are plans to improve entrepreneurial capacity of farmers, undertake market reforms, and facilitate quicker and more beneficial diffusion of appropriate innovations, including value-added technologies (Federal Ministry of Agriculture, 2012). A summary of the current policy on Agriculture is provided in table 3.

There appear to have been some gains with the Agricultural Transformation Agenda, especially in terms of increased production of cassava and rice. Nevertheless, Nigeria is still the world's second largest importer of rice, and ongoing market reforms do not seem to have, as yet, had significant impacts on the vast majority of rural farmers, in terms of access and improved income. Moreover, there is much more to do in terms of promoting the engagement of small- scale farmers in the value chains. Further research and policy evaluation are needed in these areas.

Conclusion

This review has identified a number of key developments and challenges associated with agricultural production in Nigeria. Among others, the under-cultivation of agricultural land and the problem of low yield, and the resulting challenge of food security, have been attributed to continued use of primitive implements, heavy dependence on rain-fed agriculture, the challenge of climate change, and poor farmers' lack of access to land and micro-credit. These factors have, in turn, fed, into institutional challenges including derelict infrastructure and lack of access to markets.

With respect to policy interventions, it can be seen that most of the underlying ideas in the programmes launched by successive governments were generally recycled under new nomenclatures, and they typically address, sometimes superficially, the main problems and challenges militating against agricultural productivity and rural development. However, it appears there is significant disparity between the conception of the intervention programmes and actual implementation. Two main reasons have emerged for this: the top-down approach employed by policy makers meant that little or no consultations were made with farmers, rural dwellers and other service users at the conception and design phase, such that it was difficult for farmers to take real ownership of this process and contribute their own ideas based on actual experiences on the ground. Equally important, perhaps more so, is the overarching problem of official corruption such that significantly smaller fraction of resources and funds actually reached farmers and rural dwellers at the grassroots, and infrastructural projects were poorly or tardily done, or not carried out at all. There is need for further research into the process and challenges of implementation, monitoring, and evaluation of agricultural intervention programmes.

References

Abdu, M. & Marshall, R., 1990. Agriculture and development policy: a critical review of Nigerian experience in the period up to 1985. *Journal of Rural Studies,* Volume 6, pp. 311-323.

Abila, N., 2012. Labour arrangements in cassava production in Oyo state, Nigeria. *Tropicultural,*
30(1), pp. 31-35.

Adebayo, K., 2009. Dynamics of technology adoption in rural-based cassava processing enterprises in southwest Nigeria. *International Journal of Agricultural Economics and Rural Development,* 2(1), pp.
15-25.

Adesina, A., 2014. *Building ladders out of poverty.* Lagos, Nigerian Institute of International Affairs.

Adetiloye, K. A., 2012. Agricultural financing in Nigeria: an assessment of the Agricultural Credit Guarantee Scheme Fund (ACGSF) for food security in Nigeria. *J. Economics,* 3(1), pp. 39-48.

Adesina, A. (2012). *Agricultural Transformation Agenda : Repositioning agriculture to drive Nigeria's economy* (p. 28). Abuja. Retrieved from http://www.emrc.be/documents/document/20121205120841-agri2012-special_session-tony_bello-min_agric_nigeria.pdf

Adesina, A. (2014). *Agricultural transformation agenda- 2013 report* (p. 121).

African Development Bank. (2013). *Agriculture transformation agenda support programme- phase 1* (p. 29).

Ajao, A., Olarinde, L. & Ajetumobi, J., 2005. Comparative efficiency of mechanised and non-mechanised farms in Oyo State Nigeria. *J. Hum. Ecol.,* pp. 27-30.

Akinnagbe, A. & Ajayi, A., 2010. Challenges of farmer-led extension approaches in Nigeria. *World*
Journal of Agricultural Sciences, Volume 6, pp. 353-359.

Akinro, A., Opeyemi, D. & Ologunagba, I., 2008. Climate change and environmental degradation in the Niger Delta region of Nigeria; Its vulnerability, impacts and possible mitigations. *Research Journal of Applied Sciences,* 3(3), pp. 167-173.

Anete, A. A. & Amusa, T. A., 2010. Challenges of agricultural adaptation to climate change in Nigeria- a synthesis from the literature. *Field Actions Science Reports (Online),* Volume 4, pp. 1-12.

Anyanwu, S. & Adesope, S., 2010. Low external input agriculture and rural development in Nigeria.
New York Science Journal, 3(11), pp. 65-70.

Apata, T., Apata, O., Igbalajobi, O. & Awoniyi, S., 2010. Determinants of rural poverty in Nigeria: evidence from small-holder farmers in south-western Nigeria. *Journal of Science and Technology Education Research,* September, Volume 1, pp. 85-91.

Audu, S., 2009. Gender roles in agricultural production in the middle belt region of Nigeria. *American-Eurasian Journal of Sustainable Agriculture,* 3(4), pp. 626-629.

Awoyinka, Y. A., 2009. Effect of presidential initiatives on cassava production efficiency in Oyo state, Nigeria. *Ozean Journal of Applied Sciences,* 2(2), pp. 185-193.

Damisa, M. & Yohanna, M., 2007. Role of rural women in farm management decision making process: ordered probit analysis. *World Journal of Agricultural Sciences,* 3(4), pp. 543-546.

Daneji, M., 2011. Agricultural development intervention programmes in Nigeria (1960 to date): a review. *Savannah Journal of Agriculture,* 6(1), pp. 101-107.

Davies, R., Davies, O., Inko-Tariah, M. & Bekibele, D., 2008. The mechanization of fish farms in
Rivers State, Nigeria. *World Applied Sciences Journal,* 3(6), pp. 926-929.

ECA, 2004. *Land tenure systems and their impacts on economic security and sustainable development in Africa,* Addis- Ababa: Economic Commision for Africa (ECA).

EDO, 2003. *Multi-agency patnerships for technical change in West Africa Agriculture: Nigeria case study report on rice production,* Jos, Nigeria: Overseas Development Institute.

Eze, C. C. et al., 2010. *Agricultural financing policies and rural development in Nigeria.* Edinburgh, Agricultural Economics Society.

Fajemirokun, B., 2002. *Land resource rights: issues of public participation and access to land in*
Nigeria. Cairo, Egypt, s.n.

FAO, 2005. *Contribution of farm power to small holder livelihoods in sub-saharan Africa,* Rome, Italy: Food and Agricultural Organization of the United Nations.

FAO, 2013. *Nigeria- land use and crop production data,* Rome, Italy: United Nations Food and Agricultural Organisation.

Fashola, O., Aliyu, J. & Wakatsuki, T., 2005. *Water management practices for sustainable rice production in Nigeria.* Benin City, Nigeria, Agricultural Society of Nigeria.

Federal Ministry of Agriculture, Nigeria, 2012. *Agricultural Transformation Agenda: repositioning agriculture to drive Nigeria's economy,* Abuja: Federal Government of Nigeria.

Federal Republic of Nigeria, 1978. *Land use decree.* [Online]
Available at: http://www.nigeria-law.org/Land%20Use%20Act.htm [Accessed 15 October 2013].

Idowu, O. & Ajani, Y., 2008. *Gender dimensions of agriculture, poverty, nutrition and food security in Nigeria,* Abuja, Nigeria: International Food Policy Research Institute.

IFAD, 2009a. *Agriculture in the federal republic of Nigeria,* Rome, Italy: International Fund for
Agricultural Development.

IFAD, 2009b. *Federal republic of Nigeria: country programme evaluation,* Rome, Italy: International
Fund for Agricultural Development.

Iganiga, B. & A.Asemota, 2008. The Nigerian unorganized rural financial institutions and operations:
a framework for imporved rura credit scheme in a fragile environment. *J. Soc. Sci.,* 17(1), pp. 63-71.

International Centre for Soil Fertility and Agricultural Development (IFDC), 2005. *Developing agricultural input markets in Nigeria,* Muscle Shoals, Alabama: United States Agency for International Development.

Adesina, A. (2012). *Agricultural Transformation Agenda : Repositioning agriculture to drive Nigeria's economy* (p. 28). Abuja. Retrieved from http://www.emrc.be/documents/document/20121205120841-agri2012-special_session-tony_bello-min_agric_nigeria.pdf

Adesina, A. (2014). *Agricultural transformation agenda- 2013 report* (p. 121).

African Development Bank. (2013). *Agriculture transformation agenda support programme- phase 1* (p. 29).

Kolade, O., Harpham, T., & Kibreab, G. (2014). Institutional barriers to successful innovations: Perceptions of rural farmers and key stakeholders in southwest Nigeria. *African Journal of Science, Technology, Innovation and Development,* 6(4), 339–353. doi:10.1080/20421338.2014.966039

Koyenikan, M., 2008. Issues for agricultural extension policy in Nigeria. *Journal Of AgriculturalExtension,* 12(2), pp. 52-62.

Medugu, N., 2009. *The effect of climate change in Nigeria,* Abuja: Daily Trust.

Mkpado, M. & Onuoha, R., 2008. Refined indigenous knowledge as sources of low input agricultural technologies in sub-saharan Africa: Nigerian experience. *Journal of Rural Studies,* 15(2), pp. 1-11.

Mohammed, B. & Adulquadri, A., 2012. Comparative analysis of gender involvement in agricultural production in Nigeria. *Journal of Development and Agricultural Economics,* 4(8), pp. 240-244.

National Planning Commission, Federal Republic of Nigeria, 2011. *The Transformation Agenda: summary of Federal Government's key priority policies, programmes and projects,* Abuja: State House, Federal Government of Nigeria.

Netting, R. M., 1968. *Hill farmers of Nigeria: cultural ecology of the Kofyar of the Jos Plateau.* Seattle: University of Washington Press.

Nkakini, S., Ayotamuno, M., Ogaji, S. & Probert, S., 2006. Farm mechanisation leading to more effective energy-utilizations for cassava and yam cultivations in Rivers State, Nigeria. *Applied Energy,* Volume 83, pp. 1317-135.

Nnamchi, H. & Ozor, N., 2009. *Climate change and the uncertainties facing farming communities in the middle belt region of West Africa..* Bonn, Germany, United Nations University.

Nzeadibe, T. C. et al., 2012. Indigenous innovations for climate change adaptation in the Niger Delta region of Nigeria. *Environ Dev Sustain,* Volume 14, pp. 904-914.

Nzeadibe, T., Egbule, C., N.A.Chukwuone & V.C.Agu, 2011. *Climate change awareness and adaptation in the Niger Delta Region of Nigeria.,* Nairobi: African Technology policy Network.

Obiora, C. J., 2014. Constraints to the development of technological capabilities of climate change actors in agricultural innovation system in southeast Nigeria. *Journal of Natural Sciences Research,*
4(1), pp. 67-71.

Odurukwe, S. N., Matthews-Njoku, E. C. & Ejiogu-Okereke, N., 2008. Impacts of the women-in- agriculture (WIA) extension on women's lives: implications for subsistence agricultural production of women in Imo State, Nigeria. *Livestock Research for Rural Development,* Volume 18, pp. 1-13.

Ogunlela, V. & Ogungbile, A., 2006. *Alleviating poverty in Nigeria: a challenge for the national agricultural research system. s.l., s.n.*

Ogunlela, Y. I. & Mukthar, A. A., 2009. Gender issues in agricultural and rural development in Nigeria: the role of women. *Humanity and Social Science Journal,* Volume 4, pp. 19-30.

Ogwumike, F. O., 2002. An appraisal of poverty reduction strategies in Nigeria. *CBN Economic and
Financial Review,* March, 39(4), pp. 1-17.

Olaniyi, A. M., Ayinde, M. S. & Eyitayo, O., 2007. Economic analysis of farming household's health on crop output in Kwara's state, Nigeria. *Agricultural Journal,* 2(4), pp. 478-482.

Olaoye, J. & Rotimi, A., 2010. Measurement of agricultural mechanisation index and analysis of agricultural productivity of some farm settlements in South West Nigeria. *Agricultural Engineering International: the CIGR E-Journal,* January, Volume 12, pp. 1-

21.

Olayiwola, L. M. & Adeleye, O., 2006. *Land reforms: experience from Nigeria.* Accra,

Ghana, s.n. Oloyede, J., 2008. Informal financial sector, savings mobilisation and rural

development in Nigeria:
further evidence from Ekiti State of Nigeria. *African Economic and Business Review,* 6(1), pp. 35-63.

Olujenyo, F. O., 2006. Impact of Agricultural Development Programme (ADP) on the quality of social existence of rural dwellers in developing economies: the Ondo State (Nigeria) Agricultural Development Programme experience. *International Journal of Rural Management,* 2(2), pp. 213-226.

Oluyole, K. et al., 2013. Farm labour structure and its determinants among cocoa farmers in Nigeria.
American Journal of Rural Development, 1(1), pp. 1-5.

Omobolanle, O. L., 2008. Analysis of extension activities on farmers' productivity in southwest Nigeria. *African Journal of Agricultural Research,* 3(6), pp. 469-476.

Omonona, B., 2010. *Quantitative analysis of rural poverty in Nigeria,* Abuja: International Food
Policy Research Institute.

Oni, K., 2011. *Tillage in Nigerian agriculture.* Ilorin, International Soil Tillage Research Organisation, Nigeria Branch.

Onyebinama, U., 2004. Land reform, security of tenure and environmental conservation in Nigeria.
International Journal of Agriculture and Rural Development, 5(1), pp. 86-90.

Philip, D., Nkonya, E., Pender, J. & Oni, O. A., 2009. *Constraints to increasing agricultural productivity in Nigeria: a review,* Abuja: International Food Policy and Research Institute (IFPRI).

Phillips, T. P., Taylor, D. S., Akoroda, M. O. & Sanni, L., 2004. *A cassava industrial revolution in Nigeria: the potential for a new industrial crop,* Rome, Italy: International Fund for Agricultural Development.

Sabo, E., 2006. Participatory assessment of the impact of Women in Agriculture programme of Borno
State Nigeria. *Journal of Tropical Agriculture,* pp. 52-56.

Sanni, L. et al., 2005. Value addition to cassava in Africa: challenges and opportunities. *African Crop
Science Conference Proceedings,* Volume 7, pp. 583-590.

Sanni, S. A., 2008. Animal traction: an underused low input external technology among farming communities in Kaduna State, Nigeria. *TOPICULTURA,* Volume 26, pp. 48-52.

Stone, G. D., 1998. Social organisation of agricultural labour among the Kofyar of Nigeria: an example of web-based scholarship based on an excerpt from Settlement Ecology: the social and spatial organisation of Kofyar Agriculture. *Social Science Computer Review,* 16(1), pp. 11-15.

Tripp, R., 2006. Is low-external input technology contributiing to sustainable agricultural development?. *Natural Resource Perspectives*, November, Volume 102.

Udeh, C. A., 1989. Rural development in Nigeria. *HABITAT INTL,* Volume 13, pp.

95-100. Ukeje, E., 2004. *Mordernising small-holder agriculture in Nigeria.* Pretoria,

South Africa,
Intergovernmental Group of 24.

World Bank, 2013. *Nigeria data: world development indicators,* Washington, D.C.: The World Bank
Group.

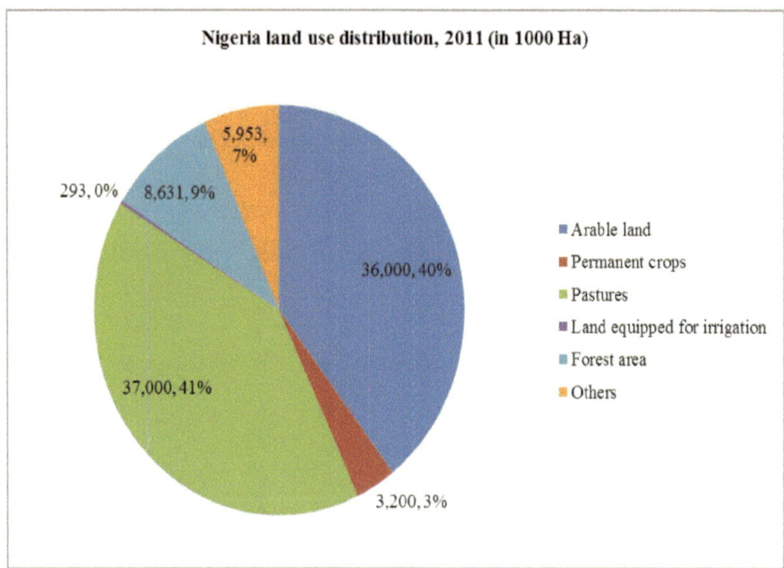

Figure 1: Nigeria's land use distribution (data source: FAO, 2013. http://faostat3.fao.org/faostat-gateway/go/to/download/R/RL/E, page accessed on 23/01/2014).

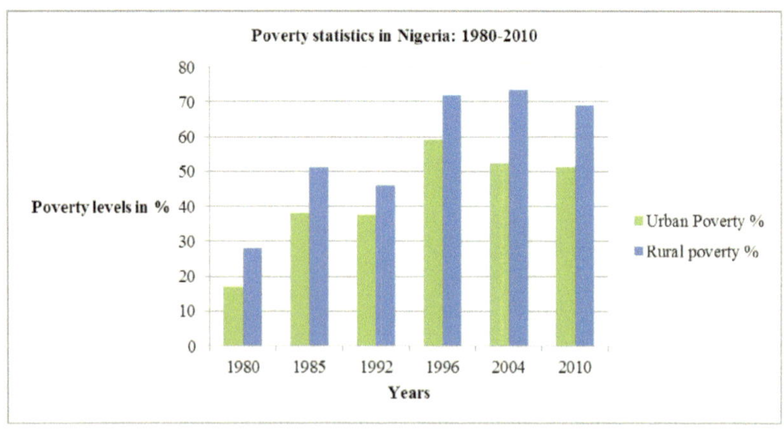

Figure 2: Poverty statistics in Nigeria: 1980-2010 (Sources: Omonona, 2010; World Bank, 2013)

Appendix B: Tables

Table 1a Statistics on production of selected crops in Nigeria between 2003 and 2012 (in metric tonnes)

Commodity	2003	2004	2005	2006	2007	2008	2009	2010	2011	2012
Cassava	36,304,000.00	38,845,000.00	41,565,000.00	45,721,000.00	43,410,000.00	44,582,000.00	36,822,250.00	42,533,180.00	52,403,455.00	54,000,000.00
Cocoa beans	385,000.00	412,000.00	441,000.00	485,000.00	360,570.00	367,020.00	363,510.00	399,200.00	400,000.00	
Cowpeas	2,459,000.00	2,631,000.00	2,815,000.00	3,040,000.00	3,000,000.00	2,916,000.00	2,371,640.00	3,368,250.00	1,860,800.00	2,500,000.00
Groundnuts	3,037,000.00	3,250,000.00	3,478,000.00	3,825,000.00	2,847,373.00	2,872,740.00	2,977,620.00	3,799,240.00	2,962,761.00	3,070,000.00
Maize	5,203,000.00	5,567,000.00	5,957,000.00	7,100,000.00	6,724,000.00	7,525,000.00	7,358,260.00	7,676,850.00	9,180,270.00	9,410,000.00
Palm oil	8,632,000.00	8,700,000.00	8,500,000.00	8,300,000.00	8,500,000.00	8,500,000.00	8,500,000.00	8,000,000.00	8,000,000.00	8,000,000.00
Rice	3,116,000.00	3,334,000.00	3,567,000.00	4,042,000.00	3,186,000.00	4,179,000.00	3,546,250.00	4,472,520.00	4,567,320.00	4,833,000.00
Yams	29,697,000.00	31,776,000.00	34,000,000.00	36,720,000.00	31,136,000.00	35,017,000.00	29,091,980.00	34,162,060.00	37,115,510.00	37,115,512.00

Data source: FAO, 2013 (page accessed on 12/10/2013).

Table 1b: Value of agricultural products using constant 2004-2006 1000 1$ (1000 Int. $)

Year	2003	2004	2005	2006	2007	2008	2009	2010	2011
Gross Production Value	31,152,616.73	33,445,104.25	35,022,650.26	36,633,507.72	34,033,980.92	36,288,928.45	31,763,926.10	35,642,129.82	35,928,195.36

Data source: FAO, 2013 (page accessed on 14/10/2013)

Table 2. Overview of agricultural intervention programmes in Nigeria (from 1960)

Year started	Programme	Description	Remarks
1960	National accelerated food production programme (NAFFP)	This programme was designed by both federal and state governments in the 1960s to accelerate the production of major staple food crops like maize, rice, guinea corn, millet, cassava, wheat and cowpea by introducing high-yielding varieties, and use of technological innovations and provision of credit (Daneji, 2011).	Research institutes were founded to help develop high-yielding varieties and work with extension agents to help farmers. One of the major institutes, The International Institute of Tropical Agriculture, Ibadan, was founded in 1967 (Abdu & Marshall, 1990).
1970s	River Basin Development Authorities (RBDAs)	These were established in the 1970s to make better and more efficient use of abundant water available for increased agricultural production beyond seasonal limitations based on rains. Multiple cropping increased, and more land areas were cultivated (Daneji, 2011; Abdu & Marshall, 1990).	Irrigation schemes were executed as significantly higher cost compared with the rest of West Africa, such that little benefit and profit were derived. In addition, many RBDAs got directly involved in agricultural production rather than facilitating opportunities for farmers (Abdu & Marshall, 1990
1972	Agricultural Development Programmes (ADPs)	This programme, which started in 1972 in the northern towns of Gombe and Gusau, focused on provision of extension services and infrastructural facilities like roads, schools and water supply to encourage farmers and increase agricultural production and profit (Olujenyo, 2006; Daneji, 2011).	This programme was assisted by the World Bank and managed under the direction of the states' ministry of agriculture and water resources. It has since been replicated in few other states, especially in the North, with moderate success.
1976	Operation Feed the Nation (OFN)	This was launched by the military government in 1976, and its principal aim is to mobilise the public to participate more in agricultural production, with incentives of input subsidies and increase in credit availability for farmers (Daneji, 2011).	Much fund was committed to publicity and media campaign, but comparatively less substantial amount to direct provision of input, credit and technical support for farmers on ground. The outcomes and results were modest (Abdu & Marshall, 1990)
1979	Green Revolution	Launched by the civilian government in 1979, this scheme provided incentives for large, medium and small scale farmers to boost production levels.	This scheme also witnessed some reorganisation of research institutes to meet the needs of farmers more efficiently.
1986	Directorate of Food, Rural and Road Infrastructure (DFRRI)	This was set up by the military government in January 1986, and the primary aim was ostensibly to incorporate a holistic measure by provision of infrastructure, technical support and other amenities to encourage farming activities and other aspects of rural life, as well as stem the pace of rural-urban migration. Reports indicate some success, with 278,526km of roads completed between 1986 and 1993 (Olayiwola & Adeleye, 2005; Udeh, 1989)	National directorates were followed by their state counterparts, with relevant divisions and department managing various aspects of mobilisation and organisation, agriculture, infrastructure, and engineering and technology, among others. However, implementation significantly failed to meet up to expectation, partly due to the rather military-style, top down approach, but mainly due to corruption on the part of the management at federal and state levels. About 5,000 rural communities were electrified, a modest success, but little compared with 100,000 communities identified (Olayiwola & Adeleye, 2005; Ogwumike, 2002)).

1992	National Agricultural Land Development Authority (NALDP)	The primary aim of this establishment was to encourage more intensified and high yielding cultivation of land available for agriculture, and also promote foreign exchange earnings by encouraging and supporting farmers to produce more than what they consume.	A central bank of Nigeria (CBN) report in 1998 indicated that this body was able to develop 18,000 ha of land, out of which about 81% were cultivated with various crops (Daneji, 2011).
2002	Presidential Initiatives	Beginning in 2002, under the civilian government, presidential initiatives were launched and implemented to focus on select agricultural produce like cassava, rice, vegetable oil, tree crops, and livestock, among others, to enhance food security at the same time as promote the crops as foreign exchange earners. Strategies employed include intensification of production to increase yield per hectare, increased and ready availability of production input and credit, identification and development of market opportunities, institutional reforms, improvement of storage and processing facilities, rehabilitation of irrigation schemes, and strengthening existing farmers' cooperatives (Philips, 2009).	Nigeria is currently the world largest producer of cassava. A 2006 evaluation of the presidential initiatives indicate the idea underlying the presidential emphasis was good and has great potential to significantly increase agricultural output and foreign exchange earnings, but lack of effective management the progress and success was not consistent and sustained. In general, for all identified crops, there was considerable increase in production output between 2001 and 2002, followed by stagnation in the next 2 years, and then a bounce again in 2005. There was no significant or steady increase, between 2001 and 2005, in the area of land cultivated for the selected crops (Philips et al, 2009)

Table 3: Key components of Nigeria's Agricultural Transformation Agenda (2011-2015)

Description	Remarks
1. Shift in conception of Agriculture as Business Enterprise, not Development Programme	The Agricultural Transformation agenda was, according to its architect, defined by a paradigm shift in terms of conception of agriculture. It was noted, with considerable justification, that previous governments conceived of agriculture as a "development project" in which government pump money with little returns to show for it, and farmers are perennially dependent on government's hand-outs(Adesina, 2012). A new conception of agriculture as business enterprise enables government to focus more on measurable impact and returns to investment, and farmers to see agricultural enterprise as the ladder to prosperity and economic mobility. Empirical tests of these claims should part of future research agenda and improvement of future interventions.
2. Re-evaluation and redesign of government's role	Historically, government has been heavily and directly involved in past interventions in the agricultural sector. Current policy seeks to reinforce new direction and thinking in which government is the facilitator and regulator, and the private sector is the direct actor in intervention programmes. For example, it is alleged that government direct procurement and distribution of subsidized fertilizers was responsible for monumental failure of that intervention. On account of endemic corruption in the public sector for four decades leading to 2011, only 11% of small-holder farmers were able to get subsidized fertilizer from government (Adesina, 2014).
3. Active integration of agriculture with the wider economic agenda and development strategy	Past intervention programmes have suffered from lack of adequate linkages and integration with other aspects of agricultural and rural development policy, on the one hand, and other aspects of the economy, on the other (African Development Bank, 2013). The current transformation agenda at least recognises the need for more effective integration, and this is why the Agricultural Transformation Agenda (ATA) is conceived as part of an overarching Economic Transformation Agenda (ETA). Thus, the ATA emphasizes the importance of market reforms and strengthening farmers' participation in high value chains. Moreover, it seeks to functionally connect with other aspects of the economy, especially services and manufacturing, and infrastructural development.

YOUR KNOWLEDGE HAS VALUE

- We will publish your bachelor's and
  master's thesis, essays and papers

- Your own eBook and book -
  sold worldwide in all relevant shops

- Earn money with each sale

Upload your text at www.GRIN.com
and publish for free